Bodies

Written by Hazel Songhurst

Wayland

CRISS X CROSS

Bodies   Fairgrounds   Light   Special Days
Boxes   Growth   Patterns   Textures
Changes Holes   Rubbish   Weather
Colours Journeys   Senses   Wheels

Picture acknowledgements

The publishers would like to thank the following for allowing their photographs to be reproduced in this book. Bruce Coleman Ltd 7 (below/Frieder Sauer), 8 (below/George McCarthy), 9 (Frieder Sauer), 10 (below/Jane Burton), 11 (above/Neville Coleman), 14 (below/Carol Hughes), 16 (below/Jose Luis Gonzalez Grande), 20 (below/Hans Reinhard), 21 (above/Jane Burton), 22 (below/Kim Taylor), 23 (Jane Burton), 25 (above/Hans Reinhard), 26 (below/Hans Reinhard), 27 (Dr Rocco Longo), 28 (below/Jane Burton); Eye Uniquitous 15 (Frank Leather), 22 (above/Mostyn); Reflections 5 (below/Jennie Woodcock), 18 (above/Jennie Woodcock), 26 (above/Jennie Woodcock); Tony Stone Worldwide 4 (Nicole Katano), 6 (Nicholas Parfitt), 11 (below/Ian Beames), 12 (G. Mortimiore), 13 (above/Michelle Garrett), 14 (above), 16 (above/ Jo Browne/Mick Smee), 17 (David Austin), 18 (below/Jean François Causse), 19 (above/Reinhard Siegel), 19 (below/Cliff Hausner), 20 (above/David Sutherland), 21 (below/James P. Rowan), 24 (Thompson & Thompson), 25 (below), 28 (above/Mervyn Rees), 29 (Peter Correz); ZEFA 5 (above), 7 (above), 10 (above), 13 (below).

First published in 1993 by
Wayland (Publishers) Ltd
61 Western Road, Hove
East Sussex BN3 1JD, England

© Copyright 1993 Wayland (Publishers) Ltd

Editor: Francesca Motisi
Designer: Jean Wheeler

Consultant: Alison Watkins is an experienced teacher with a special interest in language and reading. She has been a class teacher and the special needs coordinator for a school in Hackney. Alison wrote the notes for parents and teachers and provided the topic web.

British Library Cataloguing in Publication Data
Songhurst, Hazel.
Bodies. – (Criss Cross)
I. Title II. Series
612

ISBN 0-7502-0755-8

Typeset by DJS Fotoset Ltd, Brighton, Sussex
Printed and bound in Italy by L.E.G.O. S.p.A., Vicenza

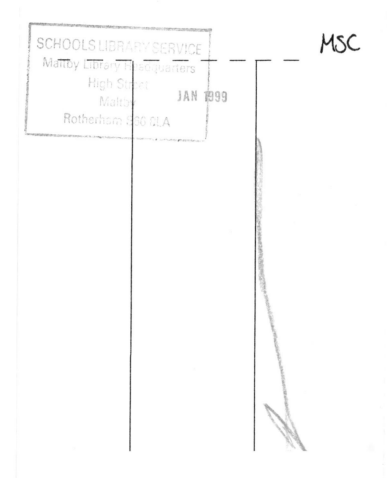
ROTHERHAM LIBRARY & INFORMATION SERVICES

Contents

Words that appear in **bold** in the text are explained in the glossary on page 32.

Clever bodies

When you run, jump and play, your body moves on its own. You don't need to tell it what to do.

When you fall down and graze your knee the skin heals quickly. Bodies can make themselves better if an accident is not too serious. This boy has broken a bone in his arm, but it will soon mend.

All living things reproduce themselves, or make babies. This woman is **pregnant**.

Big and small

Bodies come in all sizes. The largest sea animals are whales. This is a humpback whale.

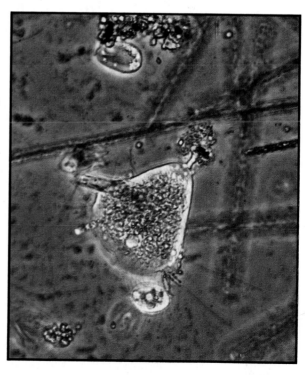

◀ The African elephant is the biggest animal that lives on land.

The smallest living bodies are nearly **invisible**! They are called protozoans and they can only be seen through a **microscope**.

Bones and cells

What do you think your body would be like without any bones?

The human skeleton has 206 bones of all shapes and sizes. Your skeleton protects and supports your body.

A bird's skeleton has hollow bones so that it is light enough to fly.

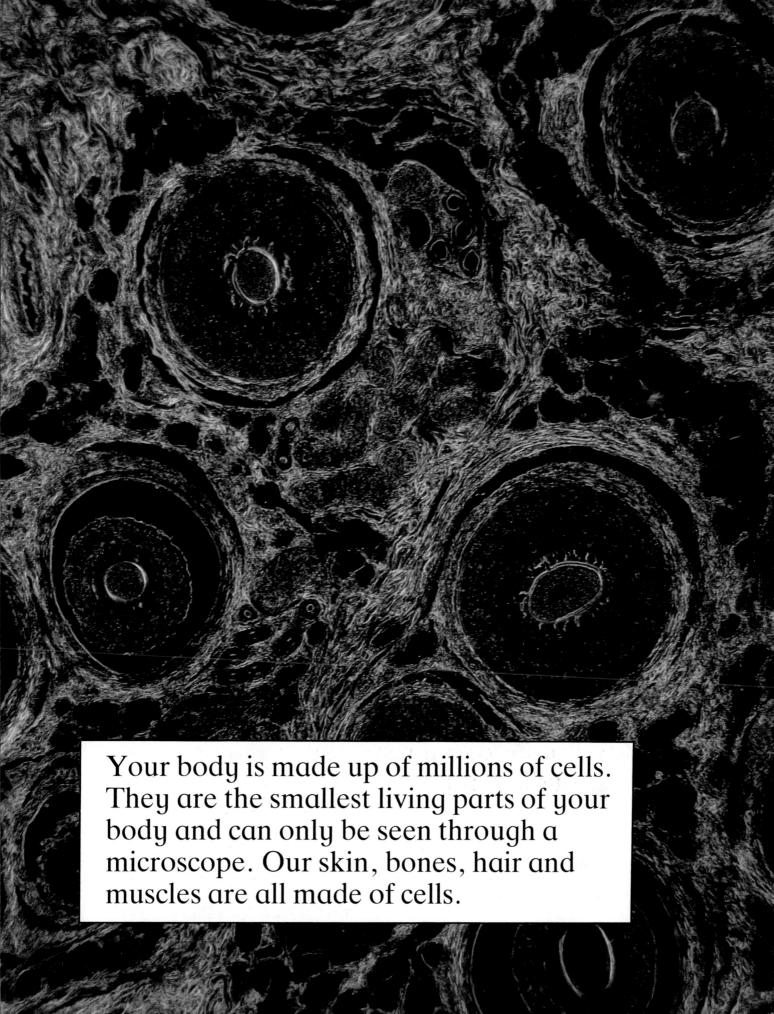

Your body is made up of millions of cells. They are the smallest living parts of your body and can only be seen through a microscope. Our skin, bones, hair and muscles are all made of cells.

Soft bodies

Worms, snails and octopuses have no skeletons. Their bodies are soft, with no bones inside them. Can you name any other soft-bodied animals?

A crab's skeleton is on the outside of its body. You can see the bony, jointed covering over its legs and the shell on its back.

This insect also has a tough **external** skeleton to protect its soft body.

Moving

Bones cannot move without muscles to pull them. Can you see the muscles standing out in the arms and legs of these runners? If you bend your arm you can make the muscle at the top bulge.

Joints are where different bones meet. They are joined together by strong **tissue**. Joints allow parts of a body to bend and make different movements.

13

This frog's long, powerful legs help it to jump far through the air. How many bends, or joints, do its legs have?

Snakes move by pushing against the ground with the strong muscles joined on to their bones. This snake is moving along by lifting itself up and then sideways.

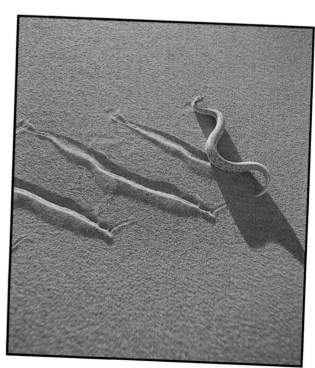

Fish push themseves along through the water by bending their tails and fins downwards then upwards.

Skin

Skin covers all bodies and protects what is inside. It is made of many layers of cells. It keeps in blood and keeps out water and germs.

Some animals have skin that does not stretch as their bodies grow. This snake is **shedding** its old, tight skin. A new skin has grown underneath.

These crocodiles have thick, bumpy skin.

Keeping warm

Bodies stay warm by trapping body heat close to the skin. When you get cold, the hairs on your skin stand out to stop the warmth from your body escaping.

Wearing thick clothes helps your body stay warm in cold weather.

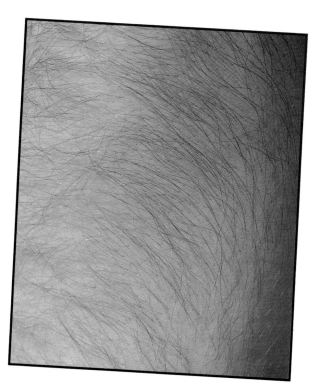

This bird has its
feathers fluffed out
to stop its body
heat escaping.

Animals that live in
cold places grow
thick, fur coats to keep
out the wet and cold.

19

Staying cool

Human bodies stay cool in hot weather by sweating. A dog loses heat through its tongue, by panting. It can also sweat through its paws and other areas of the body.

20

The ears of this desert fox give out heat from its body.

This mother and baby hippo stay cool and comfortable in a squelchy mudbath!

Keeping clean

When we have a shower we wash away the dirt and dead skin cells from our bodies. It keeps our skin healthy and makes us look and smell good!

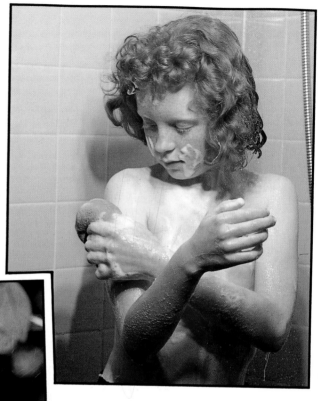

This bird is cleaning its feathers in a bird-bath. If there is no water around birds also like to clean themselves in a dust-bath.

Cats love to groom their fur.
They keep clean by licking themselves.

Food and drink

Your body needs food and drink. To stay fit and healthy it is important to eat food that is good for you, such as fruit.

Animals such as
lions and tigers
are carnivores.
This means they
need to eat meat
to keep healthy.

This huge gorilla
is vegetarian.
He prefers fruit
and leaves to
any other food.

Survival

Some bodies can do amazing things. **Mammals** can produce milk to feed their young.

A camel's body stores water and fat so that it can last for several weeks without needing to eat or drink.

Dormice **hibernate** during the winter to escape
from the cold and lack of food.

Changing

Some bodies change completely as they grow into adults. This change is called metamorphosis. Here is a butterfly coming out of its **pupa**. What was it before?

In a few weeks, tadpoles slowly change into frogs.

As we grow older our bodies change. Babies grow into children and children become adults. How have you changed since you were a baby?

Notes for parents and teachers

Science

- What makes me special? Make fingerprints (using thin paint and ink) and teethprints (bite into play dough and then fill this shape with plaster).
- Think of as many different ways as you can to move your arms and legs. Make a figure with moving arms and legs. Also the concept of movement and mobility can be explored through making puppets. (Links with design and technology, art and dance/drama here.)

Health education

- Discuss how breaking a limb, having a headache can affect our ability to perform everyday tasks.
- Explore such problems in the following ways. Ask the children to get changed for P.E. with one hand tied behind their backs, or to spend a short period of time with their ears covered. They will soon understand how difficult life can be for people with disabilities or temporary illness.
- Perhaps try working on a joint project with pupils from a special school.

Maths

- Make sets of creatures with:–
  no legs   4 legs
  2 legs   6 or more legs
- How fast am I? Use timers to find out how quick you are e.g. How many circles can you draw in one minute, skip in three minutes etc?
- Chart results and compare findings. Ask questions like does the person who can jump the best have the longest legs, or do people with the longest arms throw the furthest?

Language

- Use of fiction and non-fiction books
  Funnybones by J & A Ahlberg (Heinemann) (story tape also available)
  Happy Birthday Sam by P. Hutchins (Picture Puffin/Bodley Head)
  Tall Inside by J. Richardson (Methuen Children's Books)
  Titch by P. Hutchins (Puffin/Bodley Head).
  Also use *The Guinness Book of Records* to look at amazing bodies.

Art

- To explore the idea of skeletons outside bodies blow up a balloon and cover with papier mâché. Burst the balloon. What happens to the papier mâché? Does it collapse or keep its shape? Add wings and legs and make a creature.
- For soft-bodied creatures fill a sausage balloon with water. What happens to the balloon? Does the liquid inside keep its shape?

Music

- Song ideas:–
  One finger, one thumb, keep moving . . .
  One head, one leg, keep moving . . .
  Head, shoulders, knees and toes . . .
  The head bone's connected to the . . .

Other useful resources

Photographs	Learning/teaching packs
Films	Visits
Slides and	(museum, dentist's surgery)
filmstrips	Audio tapes
Video cassettes	Visitors and speakers
TV programmes	Posters and charts

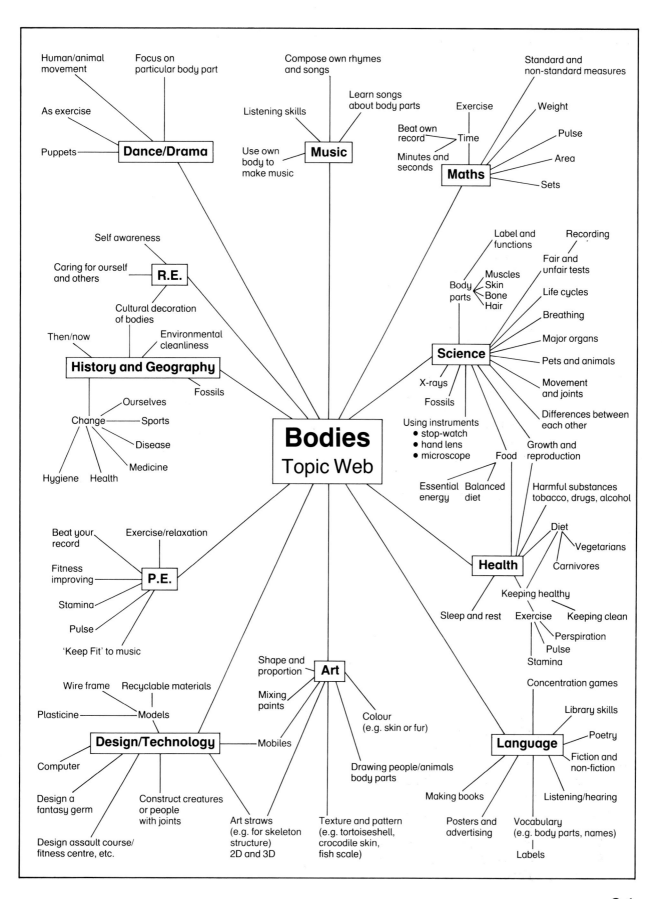

Bodies
Topic Web

Dance/Drama
- Human/animal movement
- Focus on particular body part
- As exercise
- Puppets

Music
- Compose own rhymes and songs
- Listening skills
- Learn songs about body parts
- Use own body to make music

Maths
- Standard and non-standard measures
- Weight
- Pulse
- Area
- Sets
- Time
  - Exercise
  - Beat own record
  - Minutes and seconds

R.E.
- Self awareness
- Caring for ourself and others
- Cultural decoration of bodies

History and Geography
- Then/now
- Environmental cleanliness
- Fossils
- Change
  - Ourselves
  - Sports
  - Disease
  - Medicine
  - Hygiene
  - Health

Science
- Body parts
  - Label and functions
  - Muscles
  - Skin
  - Bone
  - Hair
- Recording
- Fair and unfair tests
- Life cycles
- Breathing
- Major organs
- Pets and animals
- Movement and joints
- Differences between each other
- X-rays
- Fossils
- Using instruments
  - stop-watch
  - hand lens
  - microscope
- Food
  - Essential energy
  - Balanced diet
- Growth and reproduction

P.E.
- Beat your record
- Exercise/relaxation
- Fitness improving
- Stamina
- Pulse
- 'Keep Fit' to music

Health
- Harmful substances tobacco, drugs, alcohol
- Diet
  - Vegetarians
  - Carnivores
- Keeping healthy
  - Sleep and rest
  - Exercise
    - Perspiration
    - Pulse
    - Stamina
  - Keeping clean

Design/Technology
- Wire frame
- Recyclable materials
- Plasticine
- Models
- Computer
- Design a fantasy germ
- Construct creatures or people with joints
- Design assault course/ fitness centre, etc.
- Mobiles
- Art straws (e.g. for skeleton structure) 2D and 3D

Art
- Shape and proportion
- Mixing paints
- Colour (e.g. skin or fur)
- Drawing people/animals body parts
- Texture and pattern (e.g. tortoiseshell, crocodile skin, fish scale)

Language
- Concentration games
- Library skills
- Poetry
- Fiction and non-fiction
- Listening/hearing
- Making books
- Posters and advertising
- Vocabulary (e.g. body parts, names)
- Labels

31

Glossary

External Something that is on the outside.

Hibernate Some animals spend the winter in a very deep sleep and do not need to eat or drink during this time – they hibernate. They wake up again in the spring.

Invisible Something that is invisible cannot be seen.

Mammals Animals whose females produce milk to feed their young.

Microscope Something you use to look through to make tiny objects look bigger.

Pregnant A woman or female animal is pregnant if they are carrying an unborn baby or young animal inside them.

Pupa When a caterpillar is fully grown its skin becomes a hard, crusty shell called a pupa. Inside the pupa the caterpillar changes and grows into a butterfly. Then the butterfly hatches out of the pupa and flies away.

Shedding When you shed something you take it off.

Tissue Flesh and muscle are made of tissue.

32

Index